NULL SET

NULL
SET

POEMS

Ted Mathys

COFFEE HOUSE PRESS

MINNEAPOLIS 2015

Coffee House Press books are available to the trade through our primary distributor, Consortium Book Sales & Distribution, cbsd.com or (800) 283-3572. For personal orders, catalogs, or other information, write to: info@coffeehousepress.org.

Coffee House Press is a nonprofit literary publishing house. Support from private foundations, corporate giving programs, government programs, and generous individuals helps make the publication of our books possible. We gratefully acknowledge their support in detail in the back of this book.

Visit us at coffeehousepress.org.

LIBRARY OF CONGRESS CIP INFORMATION

Mathys, Ted, 1979–
[Poems. Selections]
Null set / Ted Mathys.
pages cm
ISBN 978-1-56689-403-6
I. Title.

PS3613.A829A6 2015
811'.6—dc23

PRINTED IN THE UNITED STATES
FIRST EDITION | FIRST PRINTING

ACKNOWLEDGMENTS

Some of these poems have appeared in the following publications: *Academy of American Poets* (www.poets.org), *Bright Pink Mosquito, Conduit, Critical Quarterly* (UK), *Denver Quarterly, Gulf Coast,* the *Massachusetts Review, Poetry Society of America* (www.poetrysociety.org), and *Portland Review.*

"Interior with Falling Bodies" was first published as a letterpress broadside, designed and printed by Elysia Mann at All Along Press.

My gratitude to the following poets for their friendship and deep engagement with this work: Mark Levine, James Galvin, Richard Kenney, Joyelle McSweeney, Donna Stonecipher, Daniel Poppick, Chris Martin, Mary Austin Speaker, Brandon Shimoda, Devin Johnston, Greg Hewett, Nathaniel Farrell, and Lisa Wells.

Rachel and Lucy, thank you for carrying me across.

CONTENTS

The Exactness   /   1

∅

Descent   /   5
Battening Song   /   7
Polyhedral   /   8
The Coast of Negative Space   /   10
Hypotenuse   /   11
Let Muddy Water Sit and It Grows Clear   /   13
Unfinished Saint Jerome   /   14
Artifact Hotel   /   16
Vikings Did Not Have Horns on Their Helmets   /   18

∅

Spurs   /   23

∅

Wildbane   /   35
Interior with Falling Bodies   /   36
Blueprint   /   37
Seven Analects   /   38
Double Exposure   /   39
Divergent Series   /   41
Brick City   /   42
The Principle of Least Animosity   /   43
Sky Correlative   /   44
Palinode   /   46

∅

All   /   51

Notes   /   77

NULL SET

THE EXACTNESS

Nothing says animal soul like the antimicrobial
proteins in tears. Nothing says regret like a pool

drained, her hushed laughter gliding its lip,
says purification like riding through the fire pit

on a horse doused with water in San Bartolomé.
The exactness is what it says it cannot be,

says what it says it cannot. Nothing rhymes
with circle, with music. Never kill a man

who says nothing, who mouths with exactness
what he says is meaningless while pointing

at the entrance to the cave. Nothing says cedar,
bluebell, solar flare. Says inversion within reason

is the story without content. Hypodermic
in clover, the exactness violates as it enters

inviolably, finds what it says cannot be.
Nothing more in the marrow than ultrasound,

tomorrow, tomorrow. The exactness was to you
as nothing is to me. Nothing says I need you

like the exactness, which sees what it claims
is invisible, claims nothing at first about

an apple, twisted off its branch,
availed to the counsel of clouds.

DESCENT

Sudden strobe
lightning lights

a stairwell you go
down, sworn

night of the Perseids'
obscure storm

into a future you
should own,

animate angles
louvered windows

you descend to shut
sore and foreign

to its warm, worn
character, her

great formal
instrument of sleep

made novel again
by light, no light, benign

light, ciphered
meteors passing

unseen, in day
as night, at least

the pear tree old
and hollow out

back is leaning,
laden, unripe.

BATTENING SONG

Doing dishtowels while waiting
to swing the axe. To sing a lullaby
about doing the honors while looking

atrocious. To cinch the lariat
around the vocal apparatus
while doing nothing but proposing

paradise. Its microclimate
and solar array. Doing a Mylar
helium balloon then reciting

the Pledge of Allegiance.
Proposing to believe. To routinize
failure into a form of hoping

to take the auspices
seriously. To follow them
out of the idiolect, motioning

toward forgiveness in that place
where it rains, but weather
is small and lenient. Along the way

to burden the sentence. To be
that burden, merging into traffic
with the precision of a bird.

POLYHEDRAL

The yellow block won't drop
through the rhombus slot
unless turned just

so, a collapse of three
dimensions into two
her small fist lacks

the precision to propose.
Months thwack wood
beneath my hand enclosing

her hand enclosing.
June won't conform to *awe,*
August to *done,* October to *on, on,*

a green pentahedron
disappears into the dark box.
She shrieks, blows raspberries.

Inevitable window. A hawk
against unadulterated sky
scanning for a kill.

I want to slide it clean
through "a hawk
against unadulterated sky

scanning for a kill"
but it doesn't fit. Perched
there on the angled branch,

a formal sound silent
beyond glass, body bound
by faces, volume enclosed.

THE COAST OF NEGATIVE SPACE

New Jersey arrives in Missouri
as a blue span in the canopy
seen from this rotting pavilion
mottled with menthol butts
when the sycamores are in leaf.

Like the mobile relation between
a thought and its rate of dispersion,
states change when I leave.
I board the train to Hoboken
in the crown of the western tree,

pass the half-remembered bank,
hospital, panhandle of sky
commanding the gap between.
Circling the roundabout fountain,
Jersey collapses in a zigzag twig.

Holland Tunnel spits me out
onto the pinnate leaf where we met.
Canal Street's mechanical pigs
flap their wings in the green stalls
we strolled. I do. The past is

a value problem in which
figure and ground compete.
Our faces near each other.
The scenario gains a squirrel
as we revert to a vase.

HYPOTENUSE

I write *three*, erase it, blow rubber
shavings from the desk. I write its glyph,
erase it, blow shavings. Then three 3s

erased, their shavings blown, persist
for the nonce, assigned to no
discrete objects I can find,

themselves objects at any rate.
To kiss, sleep, and focus we know to close
our eyes, imagine. I do, see nothing.

Traces of three 3s on a blank black plane
with their consequences and demands
attached. I have felt that way. I attach one

to a penny, a nickel, and a quarter
from the change bowl to which I four
years ago scotch-taped the quote

"The one is not." On the paper
on the desk inside the coins I write
triangle, erase it, blow shavings.

Downstairs I pour a finger of whiskey,
place kibble on kitchen linoleum,
release the dog. He trots an isosceles

that lingers once he's done,
indifferent, gone. Upstairs I rewrite
the kibble several times, erase it,

blow shavings. I attach a 3 to a film canister
full of dirt from Keats's grave that a friend
sent from Rome, a bookmark in Faulkner

down the hall, and a highball glass
in the kitchen sink downstairs.
The Dirt-Highball-Faulkner triangle

glides down, divides the house
at a 45° angle like a trowel
thrown into sod.

I write the number of letters
in sod, erase it, blow shavings, close
my eyes, go into an excavated hole

in Barberton, Ohio, which will be
a cinder block basement when my father
is done with it, winging his trowel

like a throwing knife into mud, handing me
a stake and a sledge, us squaring the footer
with a laser, orange mason twine,

crude trigonometry.
I stare down the taut hypotenuse.
It leads to an empty bottle quick with bees.

LET MUDDY WATER SIT
AND IT GROWS CLEAR

It's clear when, in membranous
        predawn blue
I enter pines, mind on
        embryo in amnion,

my tracks preceded
        by those of the dog,
his by a doe's, hers by six
        hours of snow, it's clear then

the distance between
        my affections and ability
to touch their sinuosity
        is itself a felt silence

called sun. Sun rises
        without provocation
over a frozen stream that frustrates
        reflection, but will

by the time a pulse is palpable,
        have thawed and grown
clear again, permitting me to see
        a tree surface, distort, flow.

UNFINISHED SAINT JEROME

I know why I have come to his cave.
But what to do with the lion—thorn
pulled from his paw by the half-clad

anchorite—broad skulled, tamed
to a languid S shape, futureless
and yawning at the crucifix

before the penitent's knee?
After patience, do I asphyxiate
for him a pregnant wildebeest, beat it

with the saint's hollow stone
to tenderize the meat, or leave him
unfinished furniture for the cell?

Sleeves droop over my hands.
The wall is adamant against my head.
An inscription faintly sketched.

What I have of orthodoxy I have
to lose in the hermit's shoulder
descending anatomy to whiteness,

fist paused in outline, preparing
to strike the breast it cannot strike.
Black gums, half open, unable to confess.

I cannot know. Am I adjunct
to something happening in painting
or is painting adjunct to something

happening in Rome—Rome within
the lion—caved in—incipient
and without—

ARTIFACT HOTEL

There hangs, in a hotel in Chicago,
on the thirtieth floor, behind a door
with an electric deadbolt, a photo
of the top of the hotel itself,
framed against an evacuated sky,
lightning rod piercing the gloss.
Inches below, a second photo,
this one the hotel's lower exterior,
its choir of glass. The eye strains
to suture them, not by moving
the bottom photo up to complete
the structure, but by projecting a new
nonexistent section, several floors
in height, into the gap. The hotel
spills over the top of the bottom
frame, crawls on brocade wallpaper
into the bottom of the top photo,
accommodating absence
with fluency. Behind the wall
on which the artifact hotel—taller now
than the actual—hangs, my daughter sleeps
in the other room of the suite
in a loaner crib with cold metal bars
while I watch a dramedy on mute.
How effortlessly the mind tunnels
through this wall to her, straightens
her sleep sack, smooths a curl of hair,
but I wasn't really in there, with her
inaccessible dream, in which I may
play a role, but she doesn't yet dream
in images. I am *em, oad,* speech
rudiment rather than a man in a room

inside a hotel inside itself, getting up,
tiptoeing across green carpet
to see if the floor I'm on belongs
to the bottom photo, the top, or one
I've projected in between.

VIKINGS DID NOT HAVE HORNS
ON THEIR HELMETS

The forbidden fruit was not an apple.
Poinsettias are not toxic to cats.
Marco Polo did not import pasta
from China. Diamonds are not
coal compressed. There is no access
except through aftershock, afterglow,

negation, afterlife. Abner
Doubleday did not invent baseball,
nor Marconi the radio, nor Edison
the lightbulb, nor Gore the internet.
The Pope is not sinless. Irregardless
is not *not* a word. I am an antonym

when you touch me like that.
What I mean is George Washington
did not have wooden teeth, so I live
uneasily in the law of excluded middle
where every rule has an exception.
Today the exception to the rule

that every rule has an exception
violated itself into a bright
metastasis of unfastening
while I rested my head against
aftermath. You did not run at me
yelling, "Bulls are not enraged by red!

Bananas do not grow on trees!
Toad warts are not contagious!
Sharks are not immune to cancer!"
We both know blood in the veins
is not blue, that old windows
are no thicker at the bottom

than the top. Glass does not flow
over time. Over time it's clear
this line of poetry is false.
If "this line of poetry is false" is true, then
this line of poetry is false, which means
this line of poetry is true, so

I cannot reconcile myself to finitude.
I do not enjoy being alone.
I do not have a grasp on the poetry
or the drinking. This is not going to be
transcendent. This will not blow over.
That is not what I said. What I said was

bats are not blind, and fear of death
by oscillating fan in Japan is not common.
What I mean is that eating before swimming
does not prevent cramps. Evolution
does not violate entropy. The first time
we spoke on the roof you said, "Shortly after

what came long before we met was a mutual
feeling of aftereffect." I still think of this,
how tomorrow is not the annual
Zombie March. Yesterday was not
the hundredth anniversary of the discovery
of the South Pole. Today is

not a contested topic, and in this
twos are not twos they are threes.
Meteorites are not hot when they hit
tree canopies. Men do not think of sex
every seven seconds. Fingernails do not
continue to grow after death, nor do I

know how to say no. I don't know
whether the phrase "not applicable
to itself" is applicable to itself,
but you did not deserve the panic
you have faced. You cannot gain
evidence of the raven by looking

at a pomegranate, but you can
witness a red afterimage
in which people do not die quietly
in their beds surrounded by family.
Within this afterglow we keep
trying to picture a voided shape

hovering before us at eye level
neither welcoming nor antagonistic
into which you did not stick your arm
and then did not remove it. Into which
I will not stick my arm and then not
remove it. Our arms were not,

are not, will not be permitted
inside it, nor did, are, or will we
remove, removing, remove them.
Humans do not have five senses.
Lightning can strike the same place twice.
Lightning can strike the same place twice.

SPURS

1.

Loam once
again pulls

the rug out
from under

the rug out
from under my

case in point:
I tired of

my mother
tongue so tried

and true so tried
speaking in

the foreign
tongue of a yew.

My bole of root
sounds grew

by rote
synthesis of radiant

energy until
a branch forked.

My tongue forked.
I was a liar, a snake

smelling anew
through split

tongue-tine tips
doublespeak

lying in wait.
I swallowed

a quail egg;
it broke my

foreign jaw
before I could

erect in
understory

a native liar
from litterfall.

2.

I am in the
main on the

mend I am in
Maine on

the wagon on
Katahdin in

an animal
skin I am a

pencil maker
breaking

a stolen mirror
metaphor over

the peak I
made Maine

lakes glint in
sun I broke

like a main
clause over

the forest the
page and paused

to drink from a
literal canteen.

3.　　　　　Fillet the lake
　　　　　　to find the sunset

　　　　　　faceted on water
　　　　　　split by a trolling

　　　　　　line, your wake.
　　　　　　Fillet the lake

　　　　　　trout and soak
　　　　　　flesh in a bowl

　　　　　　of water with salt
　　　　　　to reconstruct

　　　　　　sunset in an act
　　　　　　of pink astringency.

　　　　　　Fry the fillets
　　　　　　on a tripod fire.

　　　　　　From the dock
　　　　　　return the sun

　　　　　　setting in your bowl
　　　　　　back to the blackening

　　　　　　skin of the lake
　　　　　　with a watery arc

　　　　　　and splash above
　　　　　　the littoral zone

　　　　　　where minnows flit
　　　　　　in dissolving light.

4.   Lee slope
with a

bout of the
mute, I am in

montane about
to figure out

what that wild
flower goes

by, by
writing

arnica.
I am in a

parka in
arnica in no

way trying
to promote

a loss of words
at the site of seeing

a marten remove
a squirrel's face

into fact—in
fact I am

written in
rain about

to experience
it initially.

5.   In ink below
            paprika peppers

            drying in
            Andean

            sun I am
            back of this

            cactus field
            postcard,

            my course across
            its image embossed

            on your side
            under pressure

            of pen so
            run a thumb

            along the result
            of words—

            tumuli
            where I dot

            the i's I
            fell for how

            purple
            shadow fell

            on Chicoana
            as I stood in a

field of paprika
watching soil

topple cacti
as the earth

rose in large
cursive waves.

6.   Rang my
      head on

      the Bell
      Mountain Trail,

      saw stars
      rise from

      the toll
      of sundown

      on dolomite
      in a grave

      case of the
      possessive

      case. I may
      never have been

      silenced by
      Milky Way's

      cerement
      spun across

      this darkness
      had I access

      to the galaxy
      of speech I am

      assumptive in
      but apprehend

only the first
crisp day

each year
when your

breath
is visible.

WILDBANE

The ravine reveals at dusk
a grainy documentary. Amplified
by wind, silver underbellies
of leaves. Where the trail fades
to pawpaw and trillium
a tree creaks in the sway.
The black dog leaps and swallows
a black butterfly midflight.
Black pixel returned to black screen.

INTERIOR WITH FALLING BODIES

An atmospheric bacterium
forms a hailstone around itself
as a means of getting back to Earth.

Heroic in the free-fall
nucleus of the skull
my best idea was ignorant

of the structure of the storm
until broken over the sudden
hood of a Jeep. The evening news

crew weeps mysteriously.
A bomb-sniffing dog
in the international terminal

wears a vest that reads
Do Not Pet Me
unaware that nobody will.

The epiphenomenon takes
the source on a day hike
and misses the forest.

A motion light reveals my position
in the backyard of what I've been
attempting to say.

A starling mimics the ambulance.
I put on my game face
then can't get it off.

BLUEPRINT

Framing partitions in a rib cage of studs
he stumbles into my pneumatic gun—
puh-cheet!—it shoots a galvanized nail
into his lateral muscle. His mouth
dilates into a silent black disc.
The air compressor growls on.
The nail shaft is buried in flesh
but in a triumph of design, the nail head
breaks neither T-shirt nor skin; it puckers
his back fat into a funnel shape
as a black hole does gravity.
Sawdust shifting red on his shirt.
In lightless interior, the spike points
directly at the lung that houses the voice
that taught me how to hide.
Before he exhales, I extend my finger
into the funnel, past the event horizon
and press the dark spot to make stars collapse.

SEVEN ANALECTS

1. There are three topics in aesthetics: love, death, and landscape.

2. There are two topics in aesthetics: time and temporality, which are distinct in the way that *Jesus* and *Christ* are distinct.

3. There is only a monist brushing her hair in the mirror.

4. Then a hole in aesthetics where no topics exist. In the hole: *If this, then that.* But then that nullifies this. Again the hole, its worms, a landscape.

5. Is premature burial the one negative topic in the hole of aesthetics? When Houdini did it, his assistants pulled on his cramped hand the moment it broke through soil.

6. There are negative-two topics in the hole of aesthetics: mathematics and failure, which animate each other. The failure of mathematics is that it has no accepted definition, leading some to call it beautiful and pure. The mathematics of failure contains events that do not accomplish their intended purposes, thereby creating the negative lyric.

7. In the hole of aesthetics there are negative-three topics: the number seven, its notation, and *forgive them, for they know not what they do; today you will be with me in paradise; behold your son, behold your mother; why have you forsaken me; I thirst; it is finished; into your hands I commit my spirit.*

DOUBLE EXPOSURE

Before I develop the negatives
sun sears the canyon
    carbon dark. My mother leans
into a white tree, about to speak.
    Her metallic arm advances,
folded at the wrist. It belongs
    here, in the chapel of gestures.
Wind spreads her cocaine hair
    away from a finch she studies.
The bird appears confused
    by the cloud in her mouth.

/

We deal seven-card stud
    onto the side of the rowboat.
Submerged, our kitchen table
    reflects the trees ringing the pond.
She holds her cheat sheet
    in her left hand, showing
5♦ 9♣ K♥ on the oar. In the hull
    I row through the difference
between what is included
    and what belongs.

/

She emerges from a stone wall
in Carcassonne, in a denim jacket,
    red helmet balanced on her hip.
Grasses metabolize the hills.
    People bend over her motorcycle
inspecting eggs, cheeses, charcuterie.

/

Monochrome periphery. From a distance
        she leans in profile over a boulder
at the base of the waterfall rushing
        down my wedding reception, plucks
a wildflower from among the candles.

/

        In a plastic chair in sand
on an island in Lake Erie. Gulls.
        Water up to her knees. She calls this
chairing. Atop the cottage behind her
        our driveway writhes out of the chimney.
It ends in a parking cone she holds up
        like a Victrola. Whatever she said
or now says to me passes through.
        New static. Same melody.

DIVERGENT SERIES

Like crying, the desire to find common ground
between science and religion deflects attention
from the political authority of corporate executives.
Van Eyck paints all the mourners who weep
at the fringe of his crucifixion scene, out among
donkeys and plebes, a pathology of catharsis
like spiritual Kevlar. I meant to do some good
from inside the blown fuse, but confronted
personhood. The scolded elephant sobs
as apology only in the mind of the trainer.
Depending on where we drop our parentheses,
$1 - 1 + 1 - 1 + 1 - \ldots$ yields zero or one at infinity
but never both or simultaneously. Not to split hairs,
to twist them together into a thread of excess
that lacks a sum in the usual sense, in lacrimal sex,
tangible assets, the contract killing of Lazarus,
dyed between red and infrared, the adulterer
and Holy Ghost, where light becomes heat
at a wavelength the width of a needle point
we might use to sew up a burn hole in the flag.

BRICK CITY

Beer o'clock encroaches on lunch.
My reasons are as good as yours.
The rising sense that the day-
lilies will shut their faces
before sunset is ready. Access
to apocalypse is difficult for kids
so comics of zombies and I
watch a viper in the herpetarium
yawn gloriously. The hollowed-out
downtown allows the largest
coal outfit on Earth to repurpose
a library as office space, a revolt
of the referent carved in the façade:
This temple upon whose altar
is ever glowing the flame
at which patriotism is rekindled.
I idle on a lawn chair that burglars used
last night to remove the retired cop's
window-AC unit, liquidate his house
while his family slept, load electronics
into his daughter's car with keys
they lifted from her dresser.
It backfired, waking the cop.
He chased them up Wyoming Street
in boxers, gun drawn. They were fifteen
and in socks. Blood never knows
where it is in the ventricle's schedule.
Sense of relief in relief of the senses.
The fetters of inertia and then some.

THE PRINCIPLE OF LEAST ANIMOSITY

Black, jackbooted and otherwise clad in white,
slack jawed: enemy of my enemy, who for a while
slank into my idiotic alliance, why braise the whole
shank when hot marrow makes the point: a whale
shan't blow the sea unless in it, surfacing, ditto shale
shalt not resist our bulimic frack water. Enemy^2, shalt
shale plays stage a wildcatter play of *can't* v. *shan't*?
Whalebone yoked into corset: proof. Prison shank,
whole from parts, drawing blood: proof. You slank
while our enemy's yin yang beat us both slack.
White: relation is fugitive in assessments of black.

SKY CORRELATIVE

Amphoras he knows.

Bronze ingots, ivory, Canaanite jars.

In twenty years under water, my seatmate, a marine archeologist, has seen
    (he says, passing me a demibag of pretzels in the window seat)

tortoise carapaces, soapstone, an early Islamic oil lamp, hippopotamus teeth,
    ancient sex toys, orpiment, bracelets, swords, myrrh, almonds, and
    safflower all

hauled up from hulls and larders of merchant vessels, frigates, a masted
    barque, two trawlers that collided near Iceland, brig-sloops, a battle
    cruiser, Caligula's pleasure barge, boats wrecked by

squalls, tides, design error, jetty rocks, warfare, piracy, mutiny, sabotage, fire.

We're over the ocean.

As he speaks I seek these objects in clouds, work to assemble them into a
    master scene, something biblical or prurient. Content is irrelevant if I can
    find a pattern, but I can't.

My career ended in glass, he says, three tons of shattered medieval glass—
    turquoise, lavender, cobalt—we found sealed in a vessel off the coast of
    Turkey, explored

from a submersible decompression chamber without surface support, battling
    silt, salinity, wave action, poor light penetration, a slashed budget, and a
    seismic event, and when we got the cargo up we had a

jigsaw puzzle. I spent eleven years on a team that spent twenty years trying to reassemble the glass into beakers, bottles, vases, reliquaries, bowls, windows, and no one got anywhere.

He extends a cupped palm to me. Unaccountably, I take it in my hands.

One sixteen-month stretch I put together a concave section this big, he says. Nine years after I left, they discovered the vessel had been transporting scrap glass.

He retracts his hand to his tray table and rests his head on the seat back in front of him.

Over the wing the clouds are golden in impenitent sun.

PALINODE

Wake, brother, wake up,
not from what was, not from a clean
go at it, devotional and blacklisted
from the casino of punishments,
yourself having once felt Azrael's
profane wing graze your cheek

and survived. Wake, brother, not from
what used to be, an exurb commute,
actuarial exams, your deaconry
and football team that won,
season on season, even in losing
your allegiance. Not from your wife

healthy, photos of Hawaii, you with a lei
strung around your jet lag, she laughing
in the snapshot, laughing on the sofa
hunched over the album, your daughter
in diapers, lawn clippings stuck
to popsicle juice on her neck.

Wake, wake up, not from hazards
manifest, but from what had not,
could not have ever been. Wake,
dear brother, not to what should,
what might have been, not to the steady
progression of the sun along its meridian,

indiscriminate in its attribution, to each
and each, a dawn, a noon emblazoned,
then red-winged blackbirds gathering
on corn silks for decline and dusk. No,
wake, brother, to what cannot,
must not, be. Wake from the slow

delusion of reason, from the desire
to move your wife with words you know
will fail, from wanting to *be* better
to wanting to *get* better. Wake, brother,
into what cannot be, but is, powerlessness
to armor her, to find a dilated loophole

in the ruin, where stem cells will
bloom in the cerebellum, or not,
but she will appreciate you nonetheless,
where you will tell me not to puff you up
anymore, to retract the bit about strong,
patient husband and father, bullheaded

in refusal—you only ever wanted
to hear it from her, the rest of us
strangers on airwaves—not to beget
these words to you, or if I do,
to do it in writing, where you might
wake, brother, from them.

ALL

all of a piece
peculiar grace

that yet
brancht forth

—Ronald Johnson

ALL

Of my uncles, the one who ate avocados in early
fog, who took a spritzer bottle of vinegar to the pig on
the spit, who remodeled the airport on drywall stilts,
whose banjo replaced the daughter he disowned,
whose ponytail, who played chess and slept in Texas
Stadium, whose flannel and gentleness, who was opaque
as the smiley face his wife made in pancake batter in
the skillet, on whose shirt: WHEN THE GOIN' GETS
TUFF, THE TUFF GET DRINKIN', who remained anony-
mous, who put marriage on repeat, who took a knife to
his throat in bed, who threw a cigar into the ocean and
waited for waves to bring it back.

Of my jobs, sharing a ziplock of orange wedges soaked in vodka then cutting a blue spruce and dragging it to the bailer through snow, lobbying the budget chairman while averting my eyes from the stormtrooper helmet on his desk, mixing mortar and adding a squirt of New Dawn for fluffiness on trowels, assisting the assistant to the director who left an uncleared history of dirty nurses on his laptop, opening perforated baggies of broccoli and plopping a ball of whipped butter on each bisected Red Lobster potato, shredding a mining corporation's records into clear bags billowing like jellyfish, holding the gun nailer's finger safety to bounce it firing down the penciled line, going over grammar in the room with geckos before leaving to dance with the post-op coke dealer, "promoting transatlantic cooperation," asking my students to consider the retailer's inventory, and the prisoner's, mapping the garbage economy, contracting a reroof for a lover I tried to love, shingles popping off the crew's flathead shovels and sliding down the pitch as if they were scaling the fish that had swallowed us.

Of the red impressions, red sumac at the yawning mouth of the nickel mine, dashing bitters into his Manhattan, red wall I ran my hand along at Tiananmen, afterbirth in a stainless-steel span, closing my eyes and turning toward the sun, raw meat stalls at Nizamuddin's mausoleum, the campfire that exceeds its form when, days later in the shower, I wash the smoke from your hair, taillights of the car ahead on cruise control five miles per hour faster than I am, so the red eyes of the not-owl I keep chasing recede over time into dilating dark . . . *there's a blood-orange moon lying on her back with her little lava toe in the air, and she's all alone in the tarry sky, but the wee melted moon doesn't care . . .*

Of the bookmarks peeking out like periscopes, the complimentary cream one from the bookstore that I walked to the morning Thatcher died and I woke up in jail, the college-ruled index card at *Pygmy seraphs— gone astray—*, a blue ribbon sewn into the spine and dangling out the bottom as if the narrative had collapsed on a silk rat, eleven lavender Post-its fountaining out every which way in evidence of indecision, the folded grading rubric for an essay on methamphetamines, a typed 1966 checkout card: FINES TWO CENTS A DAY FOR ADULTS, ONE CENT FOR CHILDREN, the streaky, slow-exposure postcard of streetlamps rending a friend's face, the ruby one on recycled stock from the café I drove to in snow while living in the farmhouse of the painter who, before the tumor, bought the seven rooms, the silo, the granary, garden tools and the cord of wood left in the garage, the frozen lake and its dormant fish, the buried fields, the sagging hemlocks, and the unused acreage—all to acquire the drafty barn in which she painted large.

Of the reasons, good and bad, I left the church, because God is original anthropomorphism, because the youth pastor was accused of fondling boys and moved to Minneapolis, because in the fallout I told my parents I had been fondled as a boy, and though I had enjoyed the attention as a boy and didn't think much of it until the pastor moved to Minneapolis, performing damage for my poor parents at the circular table where we played euchre created the illusion of depth and lent a mythical gloss to my experience, not unlike writing it in a poem, because I didn't want my body to be a temple, as temples become ruins and subject to tourists, because the ethical trespass of the writer who perpetrates and the victim who elicits false sympathy collapse into each other in words, because the Word is not God but the residue of desire, and Protestantism lacks a certain ritual nature.

Of the fuzzy peeps she bought to raise as laying hens jabber and squeak in lidless shoeboxes next to me in the backseat, a downy wave parting at the approaching hand, turned out to be roosters that my father's crew gave three months before assembling at the stanchion in the barn, from behind the table saw I found my sight line on the process, cleavered heads in one bucket, feet in another; it must have been loud, they must have plucked them, but the spraying bird Russ threw hit the corncrib feathered, hit it quiet, made a blood willow on the slats, when he turned and caught me watching.

Of the better clichés, better late (your limit case has been put to bed and over the starless pasture the unchosen approaches, sipping from a tin cup, pulling the lever) than never, better safe than (mistaken or no, carrying her up carpeted stairs to all her contraptions where better the devil you know than) sorry, better to give (what wouldn't I give, brother, to give you time, hour of iron sun, saucepans to crash together to send a warp of sparrows into sky for her to describe again) than receive, better to light a candle than curse darkness, better seal it, a sphere around wick inside blue flame inside gold opening into the exit room where all the profanities lie.

Of a sudden face-to-face with an amateur Santa in full regalia who has run an extension cord from his garage to the intersection at which I'm in the passenger seat of Ben McCauley's black Celica waiting for the green light, window down because it's balmy for December and Santa's plugged in a string of twinkly Christmas lights, strung them up his legs and around his pillow gut into his real white beard then walked across the lawn to hand me a mini candy cane, but I don't see this, I'm looking left past Ben into the darkness that surrounds us and *HO HO* hot breath on my cheek, *Fuck Me!* I yell, shocked, punching Santa in his squishy neck; green light and Ben accelerates away toward the Wal-Mart parking lot where I lean out my window, grab a shopping cart, and roll it next to the Celica as he guns into adrenaline, takes a quick left as I let the cart go, maxed-out trembling little pathetic wheels toward a grassy hill ramp where it launches into night over corn.

Of her contractions logged on my app, sixteen hours tap at the rising pain tap at reassembled silence, duration and frequency of the last three, average duration and average frequency last hour, last six hours, glowing, consulted between counterpressure and rolling pin, how are we doing, tennis balls, what's wrong, the bags are packed, shitty that's how, you have to eat saltines, moving average moving too slow, I'm serious you're doing great, don't come in yet, they said, catalogue of helplessness next to *How to Cook Everything*, can we listen to something else where the fuck is Schlegel for the dog, it says not yet still, manipulated in the kitchen to make averages move faster, tap real, real, fake, real, fake, fake, thought we might want to look at this in the future, erased.

Of the animals I hung above your first days, a gold crane, a blue frog skewered through her nose and the purple bunny by its double tail, ears turning toward the horizon, tuned in to the yellow terrier exposing his belly to you, involuntarily pointing at your toe, boomerang fragment traveling your cornea, perk up, to hear what, perk up, a red box opening to white, and whiter twice, mad lime crane, beak caved in, my palm on your forehead as you watch the orange butterfly spin from hind wing in ambient draft, the big black-and-pink death star I hung vertically from a blade, mostly invisible to you, its line rotating against the ceiling, where is my line, what weapon does it mask, double green flower upside down from the idea of stem, does it bloom toward your face or mine?

Of my fears, not snakes, but their movement in water like worms under skin, needles sinking in, most STDs, that I am on the wrong side of Pascal's Wager, that alcoholism is inertial, the dandelion sewn into my hem is invisible, tight spaces, that I am a highly functioning pervert, being startled awake, that my hair-trigger temper will cause me to hurt my daughter in a way I can't take back, suffocation in a collapsed tunnel, unwanted pregnancy, that the law's thin film—behind which my shadow has been stripped of futurity and personal effects and left to pace—will tear open; that this is the fraudulent version of some other art I don't know, that my fears are no more original than yours, that you don't exist, the letdown of inscription, that I've already said too much.

Of my aunts, the one who forwarded warnings of internet scams, whose Jesus was the reason for the season, who packed heat at craft fairs for years and gave us peanut brittle and scratch-offs when we visited her booth, who knew northern exposure for oils of crows and barns in decay, whose laughter and Indianapolis, who did Daffy Duck with verisimilitude and bowled advanced, whose pork and beans and Oreo whip, whose nervousness and night walks, whose name I never knew, who yelled for Scotty, for Dwight, who suffered too long the plumber's tongue, in whose hand I put my hand as we walked down the driveway the night of the only lunar eclipse I've seen.

Of my fables become parables and parables fables in a chiasmus that frees them from moral and maxim, lays them out on the frozen lawn, where they look at me as I look at my hands, the left hand and the right hand are not speaking to each other, so a council of hands is convened to vote on the matter, all those in favor of the right hand holding hands with the left hand, "Aye!" all those opposed, "Nay!" the ayes have it, gavel, and when under the new resolution the right hand clasps the left and the mouths of the two hands meet, interpreting this event as a kiss, the left hand moves in for more, at which time the right hand bites the left hand's lips.

Of the essential questions: Where is love, Biscuit? What is real, Skin Horse? Are you my mother? Why is tomorrow so far away? Where is love, Biscuit? Does real happen all at once, or bit by bit? Are *you* my mother? Why is today so far away? Where is love, Biscuit? Does real hurt?

Of my obvious analogies rupture into an irradi-
ated beam, palm upturned at the cusp of ravine to jack-
light a buck, split as they enter

                    confrontation,   again
entering
              eyes—ricochet

into a black walnut tree

                        splits
itself,   them, white web
           splitting as it forms   eye-
level

charged white fog
          lighting underbrush

                that burns itself

    slow at stems

Of the seashells my mother collects from the beach near the launchpad complex, soaked in bleach and blanched in sun, wouldn't amount to much more than a wampum belt for each of three granddaughters, a Lucite box of sand dollar and sea glass, and a driftwood-and-clamshell wind chime hung from the patio enclosure, clinking in evidence of the invisible edges, driving the low-slung bridge over Indian River to the air force base at low tide for cockles, lightning whelks, paw scallops in exposed bullwhips of kelp, a kid barreling down the beach in trunks and making a question mark with a boogie board in the surf, shells chipped but permitted into her fabric bag back to the five-gallon bucket where protection sheds its grains, smooth mindlessness, sand sifting down, scooping shells up to air dry on paper towels, petal, pink, hand, known, ribbed, horn, glue gun to adorn a flowerpot for her mother, who does not remember her name.

Of the boxes in the basement below the kitchen below the room in which I sleep—stacked in a ziggurat that will not withstand the flood, manned by no priest—one marked *Kitchen # 7 / Fragile Shit* in Sharpie, an empty Two-Men-and-a-Truck teetering on a full Two-Men-and-a-Truck, spare primer, diapers and charcoal, spare paint, one with broken turntable and Art Tatum's flitting fingers trapped in solo trapped in vinyl in a paper sleeve, one with neck pillows coiling into themselves, inflatable neck pillow, fleece neck pillow, lumpy one like a bag of dicks, a box of box lids, box of smaller boxes and the stack of those collapsed, flat, no longer boxes but form prior to itself like wood about to realize this whole time it's been a violin, the buried tub of your letters I can't reach for the formless inch of water starting to reek, each letter in its original envelope, posted from one of your peripatetic outposts, sorted by year into larger manila envelopes sealed by red strings looping around red circles at the clasps, up and down in infinity shape, the year we met, year we touched, smashed the guitar, year we wrote, swam in Superior, ladled water over rocks in the sauna, havoc of the trip to the cabin, year I flew to help you get clean, cowbell clanking outside the solarium from which we watched sheep pass.

Of the first letters of the first names of the lovers
spill out all at once, resist an anagram, no longer assure
me, fail as snow

                                                        A
        K
                L
                        E
    B
                G   S

                                C

                                        L
        M
                H

                V
            Y   D
            O   F   K   R

Of my suspicions, that it is all of a piece, alone and all one, always almost gone, internally consistent in its divisibility, but going on, into algorithm, if-then, into the meadow behind the Petersons' where I found a four-leaf clover and pressed it in wax in the phone book, that it is already within me, and in a lake, catfish blubbing beneath its algal skin, and in aleph, the alphabet, that it is allergic to frames but only allows me to touch it when in them, when I am alert, without purpose and person alike, that it was in the deer tongue at salt lick, in sympathetic lightning, aligned with the effort to recall, to be awake, all told, nothing et al.

Of the evidence in metals, a doorknob comes off in my hand, aluminum whip antenna above the hatchback clinging to frayed adagio through hollows of static for seven miles then surrendering to Christian country, dead-bolt lock in Amerihost Inn, barbed wire that effectively ended the open range an artist made into a rusted lace curtain for Mayor Daley, frying pan hanging from the S-hook like a pendulum clock at rest, the steel girder on Chain of Rocks Bridge from which we watch an eagle swipe a floating fish the day Missouri injects Herbert Smulls with pentobarbital, dime rolling ahead on asphalt, arsenic in the coal-ash landfill leeching upstream, *ping ping* his crescent *ping* wrench on the engine block, key to the safe where she kept the passport but not the cash or the drugs, barrel of the rifle without an owner in the corner of the bedroom I kept locked and unheated the duration of my stay, the mailbox saluting its bent little flag, yellow leaf floating along the roof gutter in rain, a ship bell I can't see resounding in fog, a sound as circular as my wedding band.

Of the families to which this table belongs, the family of objects with legs, subsuming me, the swivel chair, the cat, hello Remi, family of all things scuffed and purchased on the cheap, family of that which has been slept on, as on the floor, on the decision whether or not to hit "Send" on my resignation, belongs to things designed for a purpose, carried upstairs, arranged near light, belongs to the family of rectangles, elongated floorboard rectangles, each brick on the façade of the bungalow through the window, closed rectangular laptop, its sleep-indicator light softly breathing on and off, each book on the shelf, and in one a physical obligation that from above seems to be a rectangle, "but seen from the side it is invisible, because it is an idea," family of things stained, stained base molding from Lumber Liquidator$, stained oval picture frame around my grandfather trying hard to look middle class in a wool suit and stained shirt, the baby's face stained after beet puree, belongs to the family of other tables, turntable, water table beneath the house, former tables I recall while writing at this one, but not the word *table,* which is itself not a member of the family of all it contains.

Of the ends agree to disagree, the end of incandescence is not the end of light but the end of this small cauldron of mood leaking over the page, the end to which I spent so much time looking at paintings that lack a beginning, middle, or end, but come entire, rush up footpaths of color and dimension, acting out how to mean but not what, the end of snow, for now, postponed by snow canons, not for good, for the suicidal enthusiasm of the poem, which could go on indefinitely but must end, a slice of toast emancipated from dough, burnt dividend of the possible condition, to contend with and pretend, mean geese all over the lawn at the end of Nana's life, their blunt refusal to be symbolic, for the end of the heyday of war toys, which engendered my first visual memory, a squadron of clean pewter fighter jets aligned on the hearth, red candles on the cake dripping wax onto the blue face of a bear.

NOTES

"Let Muddy Water Sit and It Grows Clear" takes its title from a line in Thomas Meyer's translation of the *Daode Jing*.

"Unfinished Saint Jerome" refers to Leonardo da Vinci's unfinished painting *St. Jerome in the Wilderness* (c. 1480).

"Divergent Series" draws its title from the mathematical concept of a sequence of partial sums (of fractions, for example) that does not converge toward a number as the series approaches infinity. Divergent series were formally believed to be the work of the devil.

"The Principle of Least Animosity" is a chiasmic instantiation of the SHIP / DOCK theorem, as described by mathematician Ian Stewart, in which one word becomes another word by changing one letter at a time to create a new word at each stage.

"Sky Correlative" is for Rick Kenney, with gratitude.

"Palinode" is for my brother, Stephan Mathys, and uses a rhetorical structure from William Faulkner's *Absalom, Absalom!*

The epigraph to "All" is drawn from "Ark 44, The Rod of Aaron" in Ronald Johnson's *Ark*.

In "All," page 56 contains a line from Emily Dickinson's poem "Pygmy seraphs— gone astray—"; page 66 uses language from children's books by Alyssa Satin Capucilli, P. D. Eastman, and Margery Williams Bianco; page 69 references a line in an 1871 letter from Arthur Rimbaud to Georges Izambard; and page 73 includes a quote from Ron Padgett's poem "Rectangle Obligation."

COFFEE HOUSE PRESS

The mission of Coffee House Press is to publish exciting, vital, and enduring authors of our time; to delight and inspire readers; to contribute to the cultural life of our community; and to enrich our literary heritage. By building on the best traditions of publishing and the book arts, we produce books that celebrate imagination, innovation in the craft of writing, and the many authentic voices of the American experience.

Visit us at coffeehousepress.org.

FUNDER ACKNOWLEDGMENTS

Coffee House Press is an independent, nonprofit literary publisher. All of our books, including the one in your hands, are made possible through the generous support of grants and donations from corporate giving programs, state and federal support, family foundations, and the many individuals that believe in the transformational power of literature. We receive major operating support from Amazon, the Bush Foundation, the McKnight Foundation, the National Endowment for the Arts—a federal agency, and Target. This activity is made possible by the voters of Minnesota through a Minnesota State Arts Board Operating Support grant, thanks to a legislative appropriation from the arts and cultural heritage fund.

Coffee House Press receives additional support from many anonymous donors; the Alexander Family Fund; the Archer Bondarenko Munificence Fund; the Elmer L. & Eleanor J. Andersen Foundation; the David & Mary Anderson Family Foundation; the W. & R. Bernheimer Family Foundation; the E. Thomas Binger & Rebecca Rand Fund of the Minneapolis Foundation; the Patrick & Aimee Butler Family Foundation; the Buuck Family Foundation; the Carolyn Foundation; Dorsey & Whitney Foundation; Fredrikson & Byron, P.A.; the Jerome Foundation; the Lenfestey Family Foundation; the Mead Witter Foundation; the Nash Foundation; the Rehael Fund of the Minneapolis Foundation; the Schwab Charitable Fund; Schwegman, Lundberg & Woessner, P.A.; Penguin Group; the Private Client Reserve of US Bank; vsa Minnesota for the Metropolitan Regional Arts Council; the Archie D. & Bertha H. Walker Foundation; the Wells Fargo Foundation of Minnesota; and the Woessner Freeman Family Foundation.

ART WORKS.
arts.gov

TARGET.

MINNESOTA
STATE ARTS BOARD

amazon.com

THE McKNIGHT FOUNDATION

ALLAN KORNBLUM, 1949–2014

Vision is about looking at the world and seeing not what it is,
but what it could be. Allan Kornblum's leadership and vision
created Coffee House Press. To celebrate his legacy,
every book we publish in 2015 will be in his memory.

THE PUBLISHER'S CIRCLE OF COFFEE HOUSE PRESS

Publisher's Circle members make significant contributions to Coffee House Press's annual giving campaign. Understanding that a strong financial base is necessary for the press to meet the challenges and opportunities that arise each year, this group plays a crucial part in the success of our mission.

"Coffee House Press believes that American literature should be as diverse as America itself. Known for consistently championing authors whose work challenges cultural and aesthetic norms, we believe their books deserve space in the marketplace of ideas. Publishing literature has never been an easy business, and publishing literature that truly takes risks is a cause we believe is worthy of significant support. We ask you to join us today in helping to ensure the future of Coffee House Press."

—THE PUBLISHER'S CIRCLE MEMBERS
OF COFFEE HOUSE PRESS

Publisher's Circle Members Include:

Many anonymous donors
Mr. & Mrs. Rand L. Alexander
Suzanne Allen
Patricia Beithon
Bill Berkson & Connie Lewallen
Robert & Gail Buuck
Claire Casey
Louise Copeland
Jane Dalrymple-Hollo
Mary Ebert & Paul Stembler
Chris Fischbach & Katie Dublinski
Katharine Freeman
Sally French

Jocelyn Hale & Glenn Miller
Roger Hale & Nor Hall
Jeffrey Hom
Kenneth & Susan Kahn
Kenneth Koch Literary Estate
Stephen & Isabel Keating
Allan & Cinda Kornblum
Leslie Larson Maheras
Jim & Susan Lenfestey
Sarah Lutman & Rob Rudolph
Carol & Aaron Mack
George Mack
Joshua Mack
Gillian McCain
Mary & Malcolm McDermid
Sjur Midness & Briar Andresen
Peter Nelson & Jennifer Swenson
Marc Porter & James Hennessy
E. Thomas Binger & Rebecca Rand Fund of the Minneapolis Foundation
Jeffrey Sugerman & Sarah Schultz
Nan Swid
Patricia Tilton
Stu Wilson & Melissa Barker
Warren D. Woessner & Iris C. Freeman
Margaret & Angus Wurtele

For more information about the Publisher's Circle and
other ways to support Coffee House Press books, authors,
and activities, please visit www.coffeehousepress.org/support
or contact us at: info@coffeehousepress.org.

The text of *Null Set* is set in Adobe Caslon Pro.
Composition by Bookmobile Design & Digital
Publisher Services, Minneapolis, Minnesota.
Manufactured by Versa Press on acid-free paper.

TED MATHYS is the author of two previous books of poetry, *The Spoils* and *Forge*, both from Coffee House Press. The recipient of fellowships and awards from the National Endowment for the Arts, the New York Foundation for the Arts, and the Poetry Society of America, his work has appeared in the *American Poetry Review, BOMB, Boston Review, Conjunctions,* and elsewhere. He holds an MFA from the Iowa Writers' Workshop and an MA in international environmental policy from Tufts University. He lives in Saint Louis. For more information visit www.tedmathys.com.